Instructional Technique to Develop Technical Skills, Awareness, Attitude

In Engineers and Technicians of the Third World

By

Frank Sikapa

ISBN: 1-4107-0191-3 (e-book)
ISBN: 1-4107-0192-1 (Paperback)
ISBN: 1-4107-0193-X (Dust Jacket)

This book is printed on acid free paper.

1stBooks - rev. 03/06/03

Recognition

ACKNOWLEDGEMENT

I would like to express my sincere and grateful thanks to Dr David Adams (my thesis counsellor) and chairman of the Institution of Diagnostic Engineers for his continuous encouragement. Without his effort and faith, this project might have remained another dream.

Without hesitation, I thank the late Dr R. A. Collacott for his vision in founding the Institution of Diagnostic Engineers.

An Institution of which, I am proud to be a member.

Also, many thanks to all the team members of the Royal Hospital NHS Trust, Training and Development Department; Liz Fielding, Madie Labinger, Laura Murrell, Brian Herbert and Paul Lucey for their professional help towards my acquisition of the basic and necessary management skills and vast amounts of knowledge and information. Their effort and help is greatly appreciated.

May I thank the technical staff of St Bartholomew's Renal Unit for their collective interest and support.

I thank 1st Books Library for their effort and support in publishing this book.

Thank you Lorraine Sikapa (my daughter for help in proof reading the thesis).

Table of Contents

ACKNOWLEDGEMENT III

PREFACE VII

INTRODUCTION IX

1. WHAT DO I MEAN BY 'THIRD WORLD'? 1

2. WHAT IS DIALYSIS (MY FIELD OF INTEREST) 5

3. THE ROLE OF THE TECHNICIAN/ENGINEER 7

4. INSTRUCTIONAL TECHNIQUE TO DEVELOP 13

5. LESSON PLANNING 15

6. CHARACTERISTICS 17

7. THEORETICAL APPLICATION OF TECHNIQUES............................ 21

8. MENTAL APPROACH 23

9. THE ROLE OF SELF CONFIDENCE IN PROBLEM SOLVING......... 27

10. EFFICIENCY 31

11. CUSTOMER LIAISON 37

12. TRAINING.. 39

13. WORK SHOP MANAGEMENT 47

14. VOCABULARY BUILDING.. 53

15. SERVICING.. 55

16. ORGANISATION OF MAINTANENCE............................. 61

17. PRACTICAL APPLICATION OF RULES............................. 73

18. BODY LANGUAGE... 93

19.   THREE EDUCATION DOMAINS 95

20.   LEARNING THEORIES 97

21.   MOTIVATION AND REWARD 101

22.   TRADE UNION FORMATION AND MEMBERSHIP 109

23.   BENEFITS 111

24.   DIFFICULTIES TO OVERCOME......................................113

25.   LESSONS LEARNT 117

CONCLUSION...119

APPENDIX A1 123

APPENDIX A2 126

APPENDIX B 129

APPENDIX C 130

QUESTIONS. 135

BIBLIOGRAPHY 137

Preface

The author of this project seeks to find and provide a good instructional technique to develop technical skills, awareness and positive attitude in engineers and technicians of the Third World countries like Ghana, Nigeria, Gambia, Guinea, Mali and many neighbouring and distant countries.

This may hopefully enhance relationships among all technical and engineering staff and extend such practice to other areas, also encourage the development of positive attitude towards members of the organization.

In my particular case, towards members of the dialysis team, which may include doctors, nurses, dieticians, surgeons, engineers and administrators.

The author will also like to mention other organization staff such as porters and other personnel whose endeavours and co-operation are vital and necessary for the smooth running of the organization.

Introduction

In selecting this project, I have the opportunity to examine an area in which, I believe, up until now there has been lack of emphasis; namely, incorporating the development of learning skills, the need of awareness and positive attitude into the role of the technicians and engineers in the Third World country from which I come.

We are aware that 'attitude' in any given situation is a major factor in our decision making, and judgement in performing our skilful tasks, especially when dealing with life support equipment. As such, we must be aware of that fact.

I strongly feel that any organization has the obligation to instil a favourable attitude in all members of staff.

It is commonplace discussion among some technicians and indeed among engineers that 'we are in a dead-end job'. This constitutes a negative attitude. Positive action is required.

Positive action describes the tactics, which must be employed if the citizens and Government of any country continue to ignore and disregard the demand of self-improvement.

By 'positive Action', I mean; an action to overcome fear of the masses to respond to the call of correct attitude towards serving our customers. Action to disabuse the minds of those who become negative to the course because tradition and practice.

In presenting this project, I hope that my main aim of skill development, awareness and instilling a positive attitude in technicians and engineers will be achieved.

The Third

World

1. WHAT DO I MEAN BY 'THIRD WORLD'?

There has been a lot of confusion and vague assumptions about the so-called 'Third World'.

To many, it constitutes many of the developing nations; to some who may be ignorant, it suggests the coloured or black populated areas of the world. The rest may think 'Third World' related to some uncommitted population of people who do not know whether they are coming or going, and are some kind of third and negative entity of the world.

To me, 'Third World' pertains to the developing nations who were originally committed in all respects to affiliate themselves to their western masters for many hundreds of years and have come off worse with demeaned educational standards. Help can come from a commitment from our great masters to try and enhance the skill factor, awareness and attitude in the vast arena of our educational domain.

A country like Ghana should look up to Britain for help, not because she is an isolated case, but because she is our greatest colonial power of modern times. This should apply to colonial areas everywhere. They must seek advice and help from their colonial masters such as France, Portugal, Spain and Germany, etc.

The duty or the colonial masters is to educate their respective colonies. Such a movement must be the organisation of the technical and engineering workforce

An institution could be set up which can prepare the agents of progress. They must find the most able among the youth and train their special interest including attitude, awareness, technical and scientific skills. This will help to promote continuing professional development from an early stage.

The basic driving force is economic. The colonial masses can get help from their masters by establishing an educational fund to help and encourage students of the colonies to study at home and abroad and must found schools for the dissemination of the study and development of technical skills, awareness and attitude. This may give recognition to Third World technicians and engineers internationally.

Successful campaign such as the encouragement of neighbourhood engineers and technicians, and young engineers by operating industry and institution affiliated scheme to widen the scope of engineers and technicians.

As the saying goes, 'Charity begins at home'. We must therefore endeavour to set and maintain a reasonable high standard of education and training competence and take the initiative and courage and clean up our bad attitude.

This can only enhance the economic progress of the Third World, the spirit and indigenous enterprise of the people.

Frank Sikapa

Dialysis

2.   WHAT IS DIALYSIS (MY FIELD OF INTEREST)

Dialysis is the process used in extracorporeal renal replacement therapies to remove excess water, waste products, and electrolytes from the blood. This is achieved by the method of diffusion, ultrafiltration and haemofiltration. Essentially, these terms are concerned with the movement of fluid and solutes across a membrane in an artificial kidney (the dialyzer) connected to a kidney machine, which functions as a central processing unit.

The blood is drawn from the patient's arterial vessel, passed through an artificial kidney, and returned through the vein. The correct blood flow is achieved by using a blood pump, a needle (sometimes two) and bloodlines.

A prescribed electrolyte solution (the dialysate), drawn and prepared by the kidney machine, is passed through the artificial kidney. The blood and dialysate are passed through the dialyzer in opposing direction but separated by a semi – permeable membrane. The waste product obtained across the membrane of the dialyzer is then pumped to drain.

When the solutes and fluid are successfully passed across the membrane (from blood to dialysate), by operating a pressure gradient across it, then and only then the term 'dialysis' is used.

The

Role

3.   THE ROLE OF THE TECHNICIAN/ENGINEER

Equipment management, broadly speaking falls into many categories:- mechanical, electromechanical, electromedical and pure electronics.

The equipment primarily dealt with by renal technicians and engineers include both electromechanical and electromedical.

Dialysis engineering is an extremely specialised profession.

The technical personnel's main role in any organisation is to provide technical and scientific support, and to maintain the highest engineering standards; this includes the essential provision of up-to-date testing and measuring devices.

Frank Sikapa

The need for a collaborative and co-operative culture must not be overlooked.

3.1 STRUCTURE OF A TYPICAL RENAL UNIT TEAM

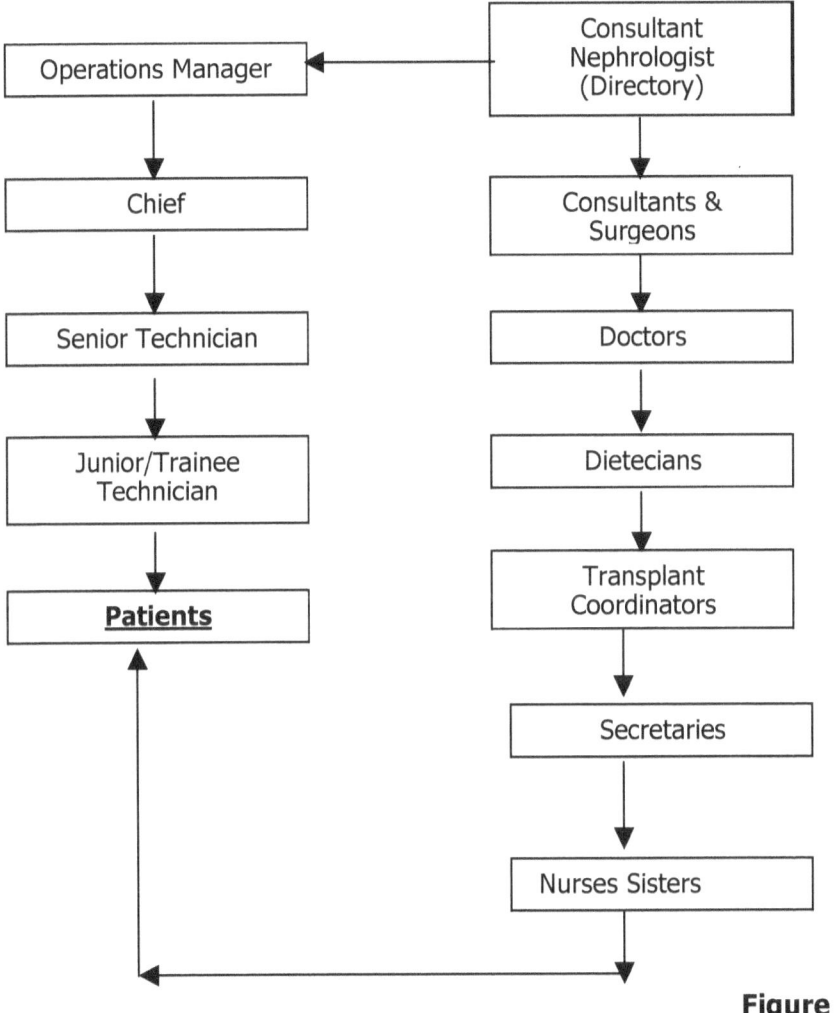

Figure 1

The above structure (which can be applied to any organisation) clearly demonstrates the wide range of disciplines with which any technician or engineer involved in the maintenance of equipment associate themselves.

3.2 EXAMPLES OF TYPICAL HUMAN BEHAVIOUR (WHICH LED TO THE THOUGHT AND PREPARATION OF THIS THESIS)

My observation spreads from Africa to Europe and my concern focuses on the following negative attitude and characteristics.

Lack of efficiency has been a problem among my compatriots.

On many occasions' bribery and corruption rear their ugly heads.

Lack of time consciousness is a problem. The pseudonym is 'African Punctuality'.

Several comments have been made regarding lack of telephone manners and response including dour countenance.

Even in Europe, it has been known that people have smoked in customer's homes.

Lack of courtesy: - Sometimes watching the television in the customer's favourite armchair and falling asleep until found out by the customer.

Feedback has intimated that some have visited customer's homes smelling of alcohol.

Lackadaisical and indifferent attitude towards work in some countries.

Favouritism and nepotism is very widely spread in the Third world.

The above negative attitude and characteristics constitute our inability to be empathetic and considerate of our customers.

Frank Sikapa

The

Technique

4. Instructional Technique To Develop

a). Technical Skills

b). Awareness

c). Attitude in engineers and technicians of the Third World countries

If we ask the question 'What is the secret of good instructional technique'? The answer could lie in good communication. All learning takes place through our natural senses, likewise with good instructions.

GOOD OBSERVATION

Immediate feedback: - This constitutes fault correction and remedy, and knowing how one is getting on in all aspects of his or her endeavours.

The correct level of advice to the trained, partly trained or untrained.

All the above are necessary in order to initiate and accomplish a successful instructional regime.

Good

Lessons

5.   LESSON PLANNING

An unplanned lesson lacks purpose and order. A successful rugby player manager has a game plan. A successful business executive has a business plan. A successful sales representative has a sales plan. Therefore we (technicians and engineers) must have and develop a learning (skill, awareness and attitude) plan in order to be successful in our technical and scientific approach in our struggle for a better working environment.

People should not be led into situations they are not totally ready for. This may cause confusion and frustration and will certainly not boost the morale of the individual concerned.

METHOD OF TRAINING

To make members feel part of the team. People must not work in isolation.

The occasional well-prepared work related presentation by employees to boost confidence and self-development. This is extremely essential for self-development in favour of self and organisation.

A well prepared session to boost and instil confidence and know how. This is to the benefit of the individual and the whole system.

Individuality

6. CHARACTERISTICS

To be encouraged: - Intrinsic values which you possess or do not, e.g. practical ability, wit and dexterity, also courtesy, empathy and consideration of others.

PRACTICE AND EMPATHY

Service engineers and technicians should not rush in and out of customers premises without due courtesy. We must aim to delight our customers.

APPERANCE AND ENTHUSIASM

Customers and patients have made comments on untidiness and incompetence of technicians, and some smelling of alcohol during service visits. This can be avoided.

CONFIDENCE AND ASSERTIVENESS

Awareness courses relating to this should be encouraged by organisation for the ultimate self-development.

CLEAR AND CONCISE VERBAL EXPRESSION

Good telephone manner is called for, especially when dealing with patients and customers in order to create a favourable interpersonal relationship.

METHODICAL AND LOGICAL

This is the most cost effective way to troubleshooting.

ASSESSMENT

Technicians performance in relation to work and loyalty to the organisation to be continually monitored. Such a practice should discipline member to always take corrective action in their day-to-day working lives.

VALUE OF PLANNING

(1) It facilitates control since it provides the objectives and standards which are the basis of any control system.

(2) It focuses attention on objectives.

(3) It brings certainty and offsets change and leads to 'pro active' management.

(4) It contributes to effective economical operation. As Koontz and O'Donnell put it: 'It substitutes joint directed effort for uncoordinated peacemeal activity, even flow of work for an even flow, and deliberate discussions for snap judgement.'

Frank Sikapa

The

Theory

7. THEORETICAL APPLICATION OF TECHNIQUES

Communication can either be verbal or written. In the technicians and engineers role I consider both to be important and necessary.

Good verbal communication among team leaders is essential to commence the basic understanding of our technical functions.

Total breakdown in verbal communication seems to create ill feeling, anger and frustration among members of the team, which can trigger the wrong attitude towards the workforce. Written communication is also vital in absence of the spoken word. The use of both form of communication can be combined for efficiency, the written communication confirming and reinforcing the verbal one.

Group discussions held first thing in the morning, or whenever convenient, have been very effective in sorting our grievances, on going work, and forward planning. **(See Fig. 6)**

Feed back from all participants is very important at this stage since it presents an overview and understanding of what activity is going on. Observation coming from the manager and organisation result in regular assessment or appraisal.

Advice from the manager to the team and vice versa should not be over looked. This may create an element of humbleness from both sides. Not many people except an overbearing personality, which can result in negative attitude towards others.

The

Approach

8. MENTAL APPROACH

How do we do this? Progress requires a lot of human effort. Individually and collectively.

According to some exponents of mental development, there are three steps to obtaining your hearts desire, including awareness and positive attitude

Desire with all your might to convert your thought or objective into fact.

Visualise, in the form of mental pictures every detail in your minds eye becoming a fact.

Sit peacefully and project that mental picture and message out ward, spreading this energy out in all direction.

To me this constitutes creative thinking and not easy to accomplish. 'Success' Edison said, 'is 1 percent inspiration and 99 percent perspiration '. Therefore by continually repeating this primary exercise, one may acquire continuity of purpose of life.

Current research shows that the left side of your brain can be in 'conflict' with the right. We should aim to address certain self-motivating difficulties by using each part of the brain to wilfully compensate for the difficulties caused by the other. A brief comparison of characteristics of left and right sides of the brain is as follows;

'LEFT' HAND BRAIN (DOMINANT SIDE); controls right side of body, verbal, logical, rational controlled, reading, writing, naming, mathematical/scientific.

'RIGHT' HAND BRAIN; controls left side of the body, non verbal, non rational emotional, intuitive, creative, face recognition, artistic, musical, songs, understanding and humour.

Progress requires human effort. Human progress depends upon individual and collective effort. Despite modern mans intellectual advancement and cultural attainment he has not yet succeeded in discarding the destructive traits of his primate forebears; because of the inborn, instructive urges, millions of humans are periodically and permanently made ill in mind and body – and purse.

This is largely due to too little reliance on self-development in terms of developing our minds, which could enhance our latent powers. Nature has endowed everyone with these latent powers capable of making them healthy, wealthy and wise.

Many erudite scientist, mathematicians, physicists, astronomers, engineers and many other professional personnel have shown the way to ultimate success through positive mental approach and development. They have worked what we ordinary men and women might term 'miracles' by demonstrating the potential powers inherent in human kind. Other men and women have even demonstrated the power to heal through positive mental domination and concentration.

Thought energy is the cause of everything. The laws of nature and mankind are activated by thought. We must therefore think positively and creatively.

We must not be to quick to criticise those who think differently to us. Shakespeare once said, 'there is nothing either good or bad, but thinking makes it so'.

Walter Russell, founder of the University of Science and Philosophy, Wanesboro, Virginia, once said; 'I believe sincerely that every man has a consummate genius within him. Some appear to have it more than othesr only because they are aware of it more than others are, and the awareness of unawareness of it is what makes each one of them into masters or holds them down to mediocrity. I believe that mediocrity is self-inflicted and that genius is self bestowed'.

He continued later by adding; 'I have found out that the real essentials of greatness in man are not written in books, nor can they be found in the schools. They are written into the inner consciousness of everyone who intensely searches for perfection in creative achievement and are understandable to such men only'.

Engineers and technicians of the Third World can emulate creativity and greatness of their colonial masters, by being aware of the fact that we also have a consummate genius within us.

In a nutshell, let us today throw away all negative thoughts which spawn unhappiness. Think positively and give to nature more than she lent.

The

Confidence

9. THE ROLE OF SELF CONFIDENCE IN PROBLEM SOLVING

The first and foremost thing is the understanding of your strength and to realise that problem solving is distinguished from decision making by the problem facing you. Since one does not even know the solution to the problem, one should consider that there is more than just 'an issue to approach' in problem solving. So initially, there is really no decision available whilst thinking of the problem.

It is even possible that the level of logic useful in approaching this problem may be limited. Both problem solving and decision making are reliant in attaining a proper understanding of one's own importance that should be a priority.

We are all very capable of finding the best possible solution to any problem. We just have to demonstrate the self-confidence to know we have the potential. Our inner self, the realm of mind that controls our personal power, should become the source of our problem solving potential.

The plan of attack to any problem-solving situation is by applying a relaxation technique. It may be wrong to approach the situation as a problem that must be solved. Tension defeats self-confidence faster than helping anything else. One must therefore view it as a problematic situation that may come up sometime along the line in future. By being at ease and prepared in advance, the solution is much easier to accomplish.

Problems are less overwhelming if they do not take us by surprise, and detected before their solutions become critical. This means engineers and technicians should develop a technique that can bring about advance warning of an impending problem. This will simplify creating a solution to it.

Self-confidence can, be enhanced by creating an early detection warning system by imagining what could happen in future. That is keeping an eye open towards possible problems areas.

Avoid snap decisions if at all possible. Instead, allow your inner mind to utilise all information you come up with and find possible solutions whilst pressure is at a low level and you can think with a clear mind.

You may not be able to anticipate every possible problem, but you will eliminate many false fears, and put yourself in an advantageous position regardig true dilemmas that may arise. This will place you in a relaxed, alert position. You will be prepared for problematic events when they do develop, and self-confidence in problem solving will be greatly enhanced. You will also create confidence among others in you.

Frank Sikapa

Proficiency

10. EFFICIENCY

It is vital for service outfits to work at high efficiency. Efficiency is a highly complex matter that relates to every aspects of our daily life. Precious resources, economic, human, spatial, are easily wasted or misused, to everyone's disadvantage in the long term. What follows is to examine several aspects of works life style, giving some constructive suggestions. Time management, personal effectiveness, meeting targets, focussing on results and opportunities are important when dealing with efficiency.

TIME MANAGEMENT DEVELOPMENT PLAN **(See Table 1)**

Should be linked with setting and meeting goals. This way, it provides the way to leverage higher achievement out of the individual and others with whom they come into contact.

Some ideas for the more effective use of time include, avoiding people at work who share your leisure time activities:- they will steal your time by talking about them. Concentrate on one thing at a time. Get up half an hour earlier each morning and go to bed half-hour earlier at night. When you want to work on something important, find somewhere inaccessible to time wasters. Write a daily work plan each morning. Tell people when they are wasting your time. Keep a time log. You can if possible work through your lunch break. One should try and review time wasted on trivial things issue an agenda for meetings, so that people have more time to prepare, holding impromptu meetings can be a time waster. Stop solving other people's problems for them.

PERSONAL EFFECTIVENESS

The factors which effect personal effectiveness are: - you, your job, the people you work with, your organisation, and external factors. An action plan is needed for your own appearance, hair, weight, clothes, awareness, attitude etc. Your workshop appearance, tidiness, walls, ceiling and decoration.

First impressions speak loudly. One of the most important aspects of appearance and one that effects the ability to work effectively in a workshop, is how the workshop is managed.

Tools must not be left on the floor and workbenches. They must be cleaned, tidied and sorted. Discard unserviceable components. Do not let service sheets and completed record forms pile up unnecessarily before filing them.

In my own workshop, we try to make good use of our main and individual diaries for better time management. To achieve this we enter key planning dates for each task. Holidays, time off are booked in advanced.

MEETING TARGETS

You need a lot of personal motivation in order to be productive. Without it, you will not only fail to achieve you goals but also you will fail at motivating others to help you achieve them. It is important to develop a self-motivating checklist.

A). Take time out to think.

B). Focus on your goals and keep difficulties in perspective.

C). See your problems as opportunities.

D). Set deadlines.

E). Work on the important not the seemingly urgent.

F). Do not pass over a difficult task – start the day with it.

A useful way of technique to change your ways of thinking about a situation, particularly when you feel demotivated is to change a negative thought pattern into a positive one. Examples are,

NEGATIVE APPROACH

I can't, I should, I hope, its not my fault, it's a problem.

POSITIVE APPROACH

I won't, I could, I know, I am responsible, its an opportunity, it's a challenge, next time I will, it's a learning experience, I know I can cope.

Thinking positively will benefit others as well as yourself including the organisation.

TIME MANAGEMENT DEVELOPMENT PLAN (Table 1)

Identify Major Time Waster	How can it be Eliminated	Target Date	Actual Date
1	1	1	1
2	2	2	2
3	3	3	3
4	4	4	4
5	5	5	5
6	6	6	6

You Won't Be Able To	So You Can	To Be Achieved By	Priority
1	1	1	1
2	2	2	2
3	3	3	3
4	4	4	4
5	5	5	5
6	6	6	6

Improve Work Environment By	Target Date	Actual Date

Frank Sikapa

The

Mediator

11. CUSTOMER LIAISON

The ability to deal with customer is of vital importance. Sometimes serving the customer can be more important than servicing the product. Those with a helpful, caring, and above all honest and sincere attitude can even turn a volatile situation or major and expensive problem to an advantage in pacifying or enhancing the relationship with the customer.

FAVOURITISM

Wise technical personal must not neglect the genuine needs of his or her customers.

NEPOTISM

We as technicians and engineers must obviate the need to be biased in favour of close relatives under any circumstances.

The above examples constitute the typical human behaviour discussed in a previous chapter;

(See, chapter 4.1). We must aim to delight our customers instead of just satisfying them.

Training

12. TRAINING

The complexity of high technology products coming from various manufacturers changing so quickly means an engineer could spend his or her whole time on training courses. In view of the cost of travelling, accommodation and 'down – time', very careful selection of training and development courses are necessary. One should make sure that the product is one likely to be used a lot or uses principles and techniques that have a wide application.

The responsibility of the management team is to identify the development needs of employees. It is up to the employee to identify the learning theories required. The flow diagram **(fig. 2),** which illustrates how employees and the manager of the organisation can take positive steps forward to identify appropriate training and development needs.

<u>THE ROLE OF THE FRONT LINE MANAGER</u>

Every front line manager has a certain amount of responsibility towards the organisation, his subordinates, himself and customers. The front line manager sometime identified as 'supervisor', 'chief technician' etc., must adopt a leadership style that integrates a high concern for people as well as for production. For the benefit of the organisation, team and self, the front line manager must be goal centred, committed and try to gain optimum results through participation. A democratic method of approach should be the preferred option.

FLOW DIAGRAM

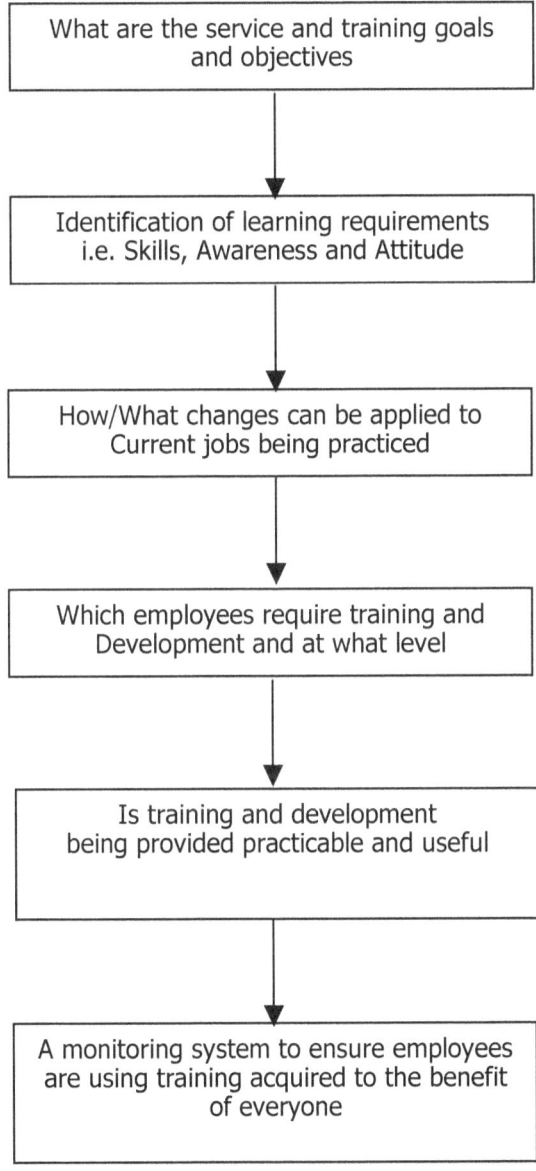

Figure 2

THE DUTIES OF A FRONT LINE MANAGER MAY INCLUDE:

A).   Planning and controlling activities to meet targets within reason, that is examine the future by drawing up an action plan. Sees that everything occurs in conformity with established rule and expressed command.

B).   Motivating subordinates to perform to the best of their ability.

C).   To delegate appropriate tasks to allow the manager to manager more effectively.

D).   To make and take decisions.

E).   To encourage ideas from the team.

F).   To recruit the right staff for the team.

A manager should be given the freedom to govern by defining the aims and expectations of his subordinates. Using the power given to him by the organisation a manager should be able to manage by co-ordinating the resources at his/her disposal to achieve a positive goal. Depending on how subordinates accept the leadership power the manager should be able to lead the group creatively to fulfil the needs of all concerned.

EQUIPMENT AND PARTS MANAGEMENT

Other duties and responsibilities of a front line manager may include: -

A). The purchasing of materials and tools plus components and spare parts.

B). The running of maintenance of stock level.

C). Providing information used in the preparation of purchasing budgets.

OBJECTIVES OF EQUIPMENT AND STOCK MANAGEMENT

A). Keep storage costs to a minimum.

B). To ensure parts are available when required.

C). To prevent deterioration, pilferage, loss, waste etc.

D). To keep accurate records and provide information to management relating to stock and stock levels.

STOCK CONTROL

OBJECTIVES: -

A). To avoid holding excessive quantities of stock. This keeps cost to a minimum.

B). To reduce the risk of running out of stock. (Stock out) this may be costly in terms of prolonging down time of equipment.

C). To bring the cost of purchasing to a minimum (procurement cost).

An effective and good stock control system requires: -

A). Accurate and up to date records.

B). Regular stocktaking.

C). Establishment of predetermined stock levels.

METHODS OF KEEPING STOCK RECORDS THAT MAY ASSIST STOCK

CONTROL ARE: -

A). **BINS SYSTEM:** - A well tried system that consists of the appropriate number of bins for individual types of stock. A working bin and a reserve bin are identified. Contents of a working bin are used up until it is empty. When the reserve bin becomes a working bin, an order is placed for the particular item or stock. The new order is then placed in the empty bin which then becomes the reserve bin and vice versa. This way, record keeping is kept to a minimum and a bin card for each stock is all that is needed.

THE IMPREST SYSTEM

This requires a pre printed stock sheet for each spare part item and recording the correct stock level at regular intervals the quantity of each item in stock is recorded and compared with normal stock level. The difference is recorded and if the amount in stock is less than the normal stock level, then the quantity required is recorded on a stock sheet which is then signed, dated and referred to the purchasing department as a purchasing requisition. This method insures that purchases are based on accurate, current stock levels and requirements. It also keeps stores to a minimum. Where such an operation is centrally organised, it acts as a formal purchase requisition procedure.

CYCLICAL REORDERING

The system involves making purchase orders for standard quantities of stock at regular intervals. This method is very simple to operate, but runs the risk of a stock – out or over stocking. It's useful when consumption of a material is constant.

THE OBJECTIVES OF STOCK TAKING

A).   Ensure the system is reliable and meets service needs.

B).   Detect errors and expedite their rectification.

C).   Check for pilferage.

D).   Provide information on stock quantities and values in report accounts of financial statements.

E).   Identify obsoletes and slow moving stock.

The

Workshop

13. WORK SHOP MANAGEMENT

Besides financial matters, politics and management, there is a more tangible side to the service operations. The lay out, facilities, tidiness and organisation of the workshop are very important factors.

With an efficient and courteous filtering system, minimising of the interruption a workshop engineer and technician face can be avoided. This contributes to a high degree the total efficiency of the system.

Finance is regarded as one of the most crucial aspects of workshop management. It is therefore important to work within your financial constraint.

The lay out should be simple and cost effective. The construction of a simple multipurpose service trolley, with suggested dimensions, which can dock into a bench console with shelves, can be quite practical. Of course this depends on the complexity of the equipment being serviced.

(See Fig. 3).

The advantage of this simple lay out is that easy access is provided to equipment being serviced or repaired. The equipment can stay on the trolley throughout repair, soak testing and transportation to and from the customer's car.

For bench lighting, a large circular fluorescent type with a central magnifier and an articulate cantilever arm is favoured by most technicians.

When dismantling any equipment, put fixing screws into a segmented tray with a compartment clearly marked top case, bottom case, back, PCB deck etc. or leave them in a progressive line as you dismantle them. This avoids loosing them or misplacing them during reassemble.

Major test equipment such as an oscilloscope, frequency meters, power supply units can be left on a trolley and made accessible to all technicians and engineers for them to use as necessary.

For safety reasons it is advisable not to isolate engineers. If at all possible the room should be big enough to accommodate all technicians or at least within shouting distance of each other.

Diagnostic engineering has a primary responsibility for safety and there should be a programme designed to reduce accidents. This must be based on accurate records and incident levels reviewed periodically.

Some money should be invested in keeping stock of important spare parts for repairing and troubleshooting purposes. Individual engineers should not be allowed to order spares at will. Training should be given to one responsible person in charge of general components, with one-off specials for particular jobs being ordered by this technician concerned.

The key to profitable servicing may lie in the fast and correct fault diagnosis. This can be achieved through experience and a good working knowledge of the equipment, availability of good service manuals and circuit diagrams, and adequate and accurate test equipment. You also need a good working environment and good

communication with manufacturer and customers.

<u>Methodical and logical approach to fault finding is essential</u>

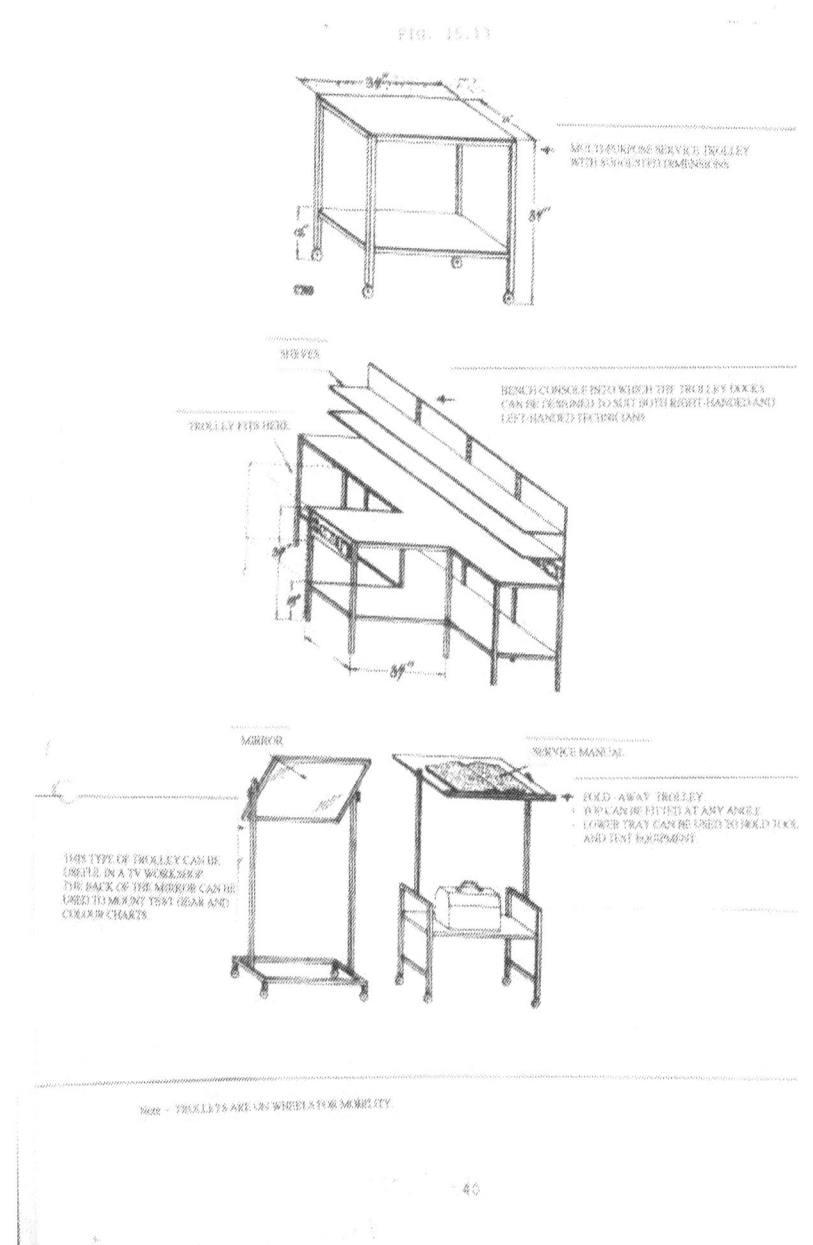

Figure 3

Frank Sikapa

Your

Vocabulary

14. VOCABULARY BUILDING

Words are tools of thought and communication. It is therefore clear that words are an essential part of all our association with others including occupational advancement.

As an African, I am aware of the lack of vocabulary when it comes to finding descriptive words, definitions for technical and scientific objects. This is difficult in a foreign language like English let alone in a n under developed African language.

Words form the symbols of ideas and accurate measurement of our mental equipment. The developed (Western) world can use the right word at the right time and has definite advantage over my compatriots who are less well equipped.

The answer to this difficulty and lack of awareness in this matter may lie in the development of our African language to accurately describe and be able to define all technical and scientific jargon. This practice will not be found easy since there are many different languages and dialects within a radius of a few miles with the African continent. The solution may lie in encouraging one main language like Swahili or any other convenient language. This will be a giant step towards success.

Africa must no longer be considered as a 'dark continent'. A lot has been written and known about this so-called 'dark continent'. Even if it was, we must realise and understand that 'the brightest light may break from the darkest sky'. (Cheiro).

It pays to enrich your word power. This may improve your communication and participation.

The

Service

15. SERVICING

The service and maintenance industry has gone through a period of rapid and radical change. This change concerns technology, scale, economies and organisation

Technology used in the product we are supposed to service gets more complex by the month. One effect of this is that unless a fault is simple, and the increase in reliability of modern equipment, means that there are fewer and fewer of this type, the difference between the cost of repairing a product and its initial cost is rapidly diminishing. Thus the unit of what is economical feasible in the way of repairs is drawing in. The rapidly increasing cost of skilled labour and the competition amongst employers for qualified people accentuates the problem.

Servicing sometimes involves efficient troubleshooting to the highest degree. The keen competition in servicing today demands prompt diagnosis of the fault and quick replacement of any faulty component parts. This may be the only cost effective way.

Servicing and troubleshooting consist of sections and orderly examination of key test points leading to a logical deduction of the cause of the symptom. This requires a study of the symptom it self, taking into consideration the frequency of the various types of failure, the reliability of test results and the practicality of using certain kinds of technical aids for testing purposes. In most troubleshooting exercise, previous knowledge and experience count a great deal. It becomes easier when stock faults are successfully accumulated.

By using reliable service charts supplied by the manufacturer, circuits, be it electronic, electrical or plumbing, whole sections can be isolated with minimum tests and efforts.

A servicing chart, which summarises the isolation procedure, is extremely important. Efficient troubleshooting procedure may include voltage measurement, resistant measurement, signal injection, the use of an oscilloscope for waveform tracing and the substitution of electrical or mechanical parts.

WORKING ON EQUIPMENT

One must equip oneself with a good selection of tools of convenient shape and size. All servicing tools must be kept in good condition. Do not forget to make a proper sketch of the circuit when several leads are disconnected at the same time in removing a part for replacement, otherwise your simple repair job will be your toughest. Take safety precautions.

Resist the temptation and impulse to adjust vital analogue controls. It is very seldom that any piece of equipment can be repaired by adjusting non-customer controls. Remember also that it is virtually impossible for any equipment to totally fail because some controls have misadjusted themselves. This might not be the case where digital circuits are used. Spikes and transient fluctuations in current and voltage may cause some calibrations to go back to default (i.e. factory presets) and complete calibration from first principle may be needed.

In many cases simple voltage checks can isolate whole sections. The very simple circuit below demonstrates how one can very quickly trace faults towards 'B+'. This is a simple technique used by all technicians after measuring for a possible voltage at some point and not finding it there. The probe is moved across each point in series with the point, moving towards 'B+'. If one of the components in this series line is faulty, the voltage appears on the 'B+' side of the component.

When for example 'C' is shorted the voltage will appear at the 'B+' end of R1. Sometimes it will be better to confirm this by removing one leg of the suspected component and measuring its resistance, inductance or capacitance as the case may be.

(See Fig.4)

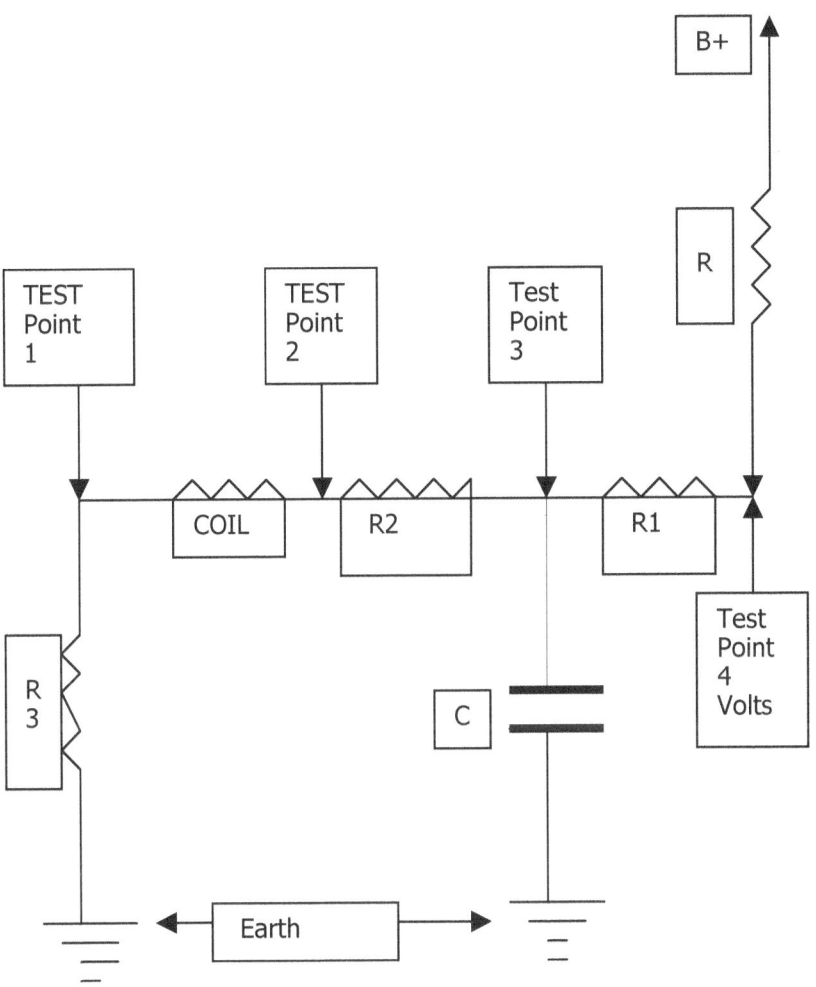

Fig. 4

Frank Sikapa

Maintenance

16. ORGANISATION OF MAINTANENCE

As mentioned in chapter 3, the task of technicians and engineers is to study and plan for a wide range of equipment adapted to the function of important needs. Also to make sure that the operation of such equipment is adequate in order to fulfil the correct function.

In order to achieve this we must ensure proper maintenance of such equipment at all levels. **See Fig. 4**

The appointment of a manager **(See Appendix 'A 1')** responsible for all the maintenance levels detailed in **Fig. 5 & 6** is recommended.

Management should be kept informed about maintenance costs, the cost of breakdown, effectiveness of the system and relevant information about when the equipment should be replaced.

The objectives of a good maintenance regime is obvious.

1.   To extend useful life of equipment.

2. Ensure operational readiness of equipment.

3. Ensure safety of personnel using the facility.

4. Ensure maximum returns of equipment.

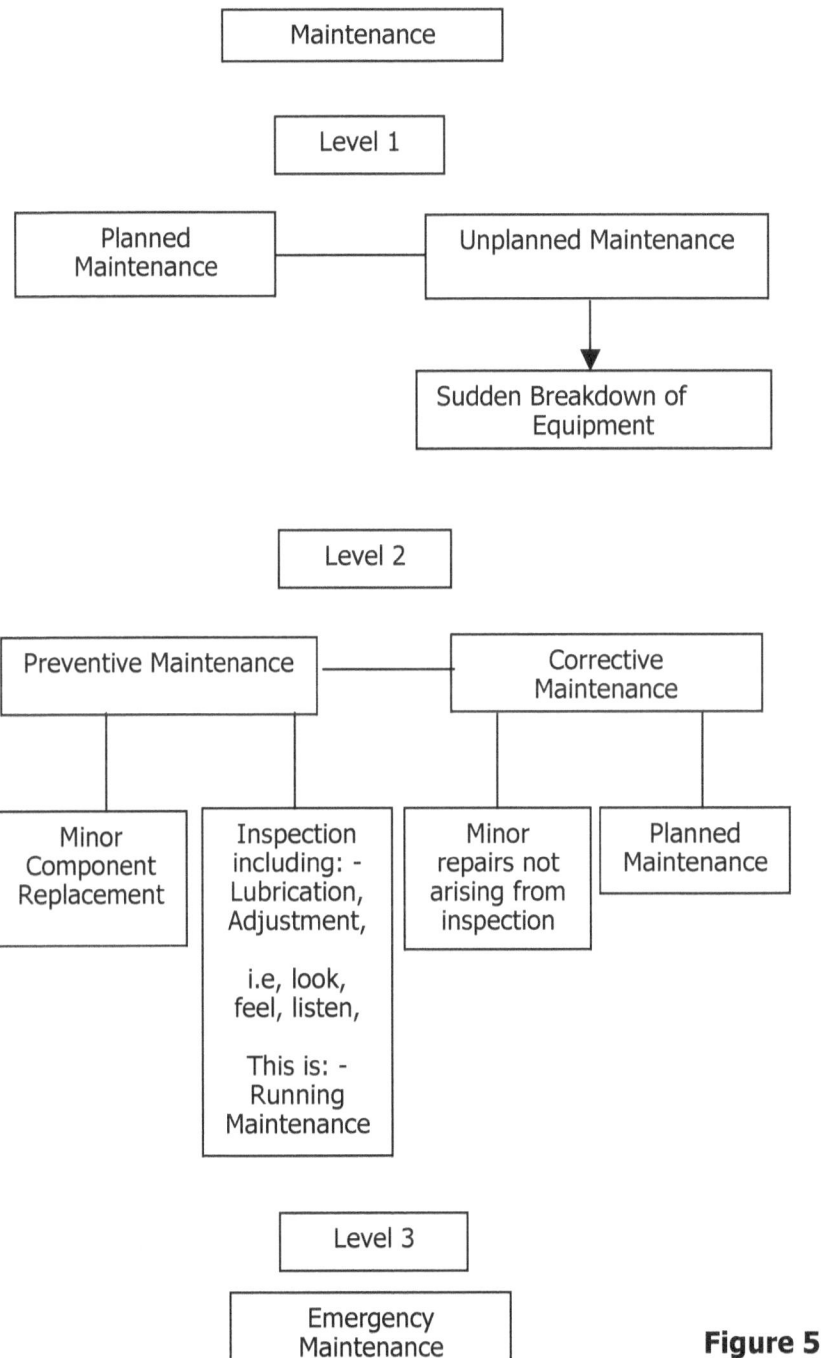

Figure 5

IN CONCLUSION,

Prepare, a:

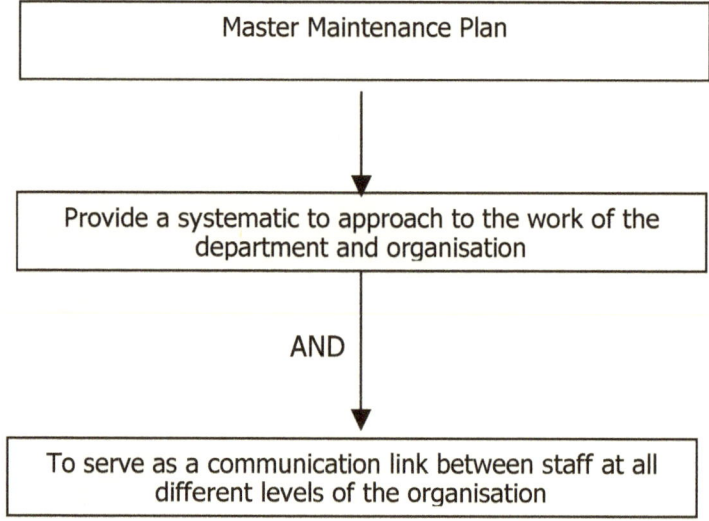

Figure 6

SUPPLIES PROCEDURE FOR USER DEPARTMENTS OF THE

ORGANISATION

Goods and services in many major organisations such as, the National Health Service, for official use may only be obtained through an area supplies division, in accordance with the standing orders and supplies policy of the area authority or organisation. Any other method used will be in contravention of this. The soul purpose of this progressive step is to centralise purchasing, storage, and distribution so that certain goods and services are obtained via the area supplies division.

ROLL OF SUPPLIES DIVISION

The supplies division aims to provide the necessary information service to all users on equipment currently available. Initial request for information should be made to the supplies division, where possible, and quotations for goods and services should not be sought by users. As often this may pre-empt the supplies division role, and duplicate effort. All request for goods and services must be channelled through a co-ordinator.

AUTHORISATION OF REQUESTS

The district supplies co-ordinator will act upon a correctly authorised requisition or order request, that is, one which is signed by the person who holds the budget for the requested item, or his designated officer. On the advice of the finance department and in consultation with heads of departments, the amount of money to be allocated to each budget is determined. Every budget holder is given an expenditure code, against which they are authorised to request a purchase to be made, and whenever possible, the financial coding must be annotated on the order request form.

ACTIONING REQUESTS

Goods requested will either be issued from stock on a scheduled delivery service to various departments, or if not available from stock, will be purchased from a reputable supplier, against a negotiated contract, or if over a certain value, by competitive quotation. Purchasing procedures in a large organisation such as the National Health Service reflect public accountability and there are limited circumstances in the area health authority standing orders when quotation procedures may be waived. In addition to accounting for expenditure of public monies, there are other requirements pertaining to safety of all customers and staff, which must be met. The working environment must be free from hazards as far as possible and this includes the equipment and sundries necessary to meet customers and staff needs. Before a purchase is made or effected, the following points must be clarified.

A. Does it meet flammability standards?

Does it meet with the electrical safety standards laid down by the organisation and other regulations?

B. Is it likely to cause a potential hazard to customers and staff?

C. Does it meet the requirements of The Health and Safety at Work Act?

D. Has a hazard warning ever been issued?

SUPPLY OF ELECTRICAL AND ELECTRO-MEDICAL

EQUIPMENT

The Department of Health has specified certain standards to which electromedical equipment and electrically operated equipment purchase for the NHS must comply. In order to meet these standards a questionnaire must be completed by potential suppliers for every piece of equipment before a purchase may be considered. In addition, the procedure must also be applied to electro-medical or electrically operated laboratory equipment supplied on hire, lease or loan, as well as to purchase a few pieces of equipment. The purpose of the questionnaire is to enable the authority to ensure not only that the equipment concerned will meet the required technical and safety standard, but also that satisfactory back up service and spare will be available.

Each completed questionnaire is examined by the Department of Medical Electronics or equivalent, and if found to be satisfactory, a purchase will be made against an authorised request. Those items which are found not to conform to standards must be identified and modified by the supplier to the required standard before a purchase may be considered. In the event that an acceptable standard cannot be reached, alternative equipment should be considered. On receipt of the item, The Department of Medical Electronics or its equivalent will carry out an 'acceptance test' **(See Tables 1 & 2)** and until such testing is carried out the equipment must not be used. This procedure will inevitably take longer than hitherto; it would therefore be to the advantage of the requesting department to notify the supplies division as soon as possible of any likely forthcoming request for such equipment.

DISPOSAL OF SURPLUS EQUIPMENT INCLUDING SCRAP

The district supplies co-ordinator is responsible for disposal of surplus equipment and scrap. In the case of surplus equipment the line manager may be able to recommend a transfer of equipment to another department within the hospital or organisation. If there is no such demand, details of item/s should be forwarded to the district supplies co-ordinator so that its availability may be circulated to other departments or organisations in the district, area or region.

Items suitable only as scrap are disposed of by competitive bidding, and in the first instance, notification should be made to the supplies co-ordinator. Monies obtained from the sale of surplus equipment will normally be credited to the district and not the budget holder requesting the sale, although in some cases prior arrangements may be made with the finance department to credit the budget of the requester.

EQUIPMENT MAINTENANCE ARRANGEMENT

The head of the department or front line manager should ensure that routine maintenance and inspection of equipment is carried out 'in house' or by specialist contractors. Subject to the head of departments approval, maintenance contracts will be raised by the District Supplies Unit, and in addition, those request for repairs (other than works maintenance) not covered by a maintenance contract should be channelled through the District Supplies Unit.

REPLACEMENT OF OUT-OF-DATE EQUIPMENT

For the reason discussed in **(chapter 16)** it is very important that user departments submit their requests for item/s replacement in good time to enable effective procedures to be under taken.

Having accepted the equipment after a successful demonstration by the supplier, the line manager has to produce an indemnity form to be completed and signed by the supplier or his/her representative. Only then can the equipment be used for clinical trials or other appropriate usage.

See a) Form of Indemnity. b) The Schedule. **(Appendix C)**

Frank Sikapa

The

Rules

17. PRACTICAL APPLICATION OF RULES

COURTESY (Practice and empathy)

Technicians will be committed to promoting a good working relationship among all members of staff and be tactful towards patients and customers.

Constraint will be exercised; people will avoid smoking in patients or customers homes.

Customers and patients including members of staff will be approached in a courteous manner.

Patients and customers rooms are to be left in a tidy and clean state; i.e. in the condition which they were found.

Customers confidentiality to be of prime importance and technicians to be made aware of this.

Technicians confidentiality also to be of prime importance; their private telephone numbers must not be given to patients or customers without the appropriate consent of individual technicians.

Technicians and engineers will avoid loitering in patients or customers homes.

Self commitment and consistency is necessary to co-operating and working as a team.

SAFETY AND PRACTICE

Technical and engineering staff may obviate the necessity for uncontrolled technical adjustments by customers and patients on all life support machines, including ancillary equipment. This is to be regarded as a safety factor. **See Appendix 'B'**

Junior technicians will seek designated authority before performing any adjustments which may effect the welfare of the patient or customer.

Six-monthly planned preventative maintenance will be carried out by technicians in line with manufacturers specifications and up to

British Standard (BS5750), see Appendix 'B' (THE SAFETY PHILOSOPHY).

With dialysis machines, routine decalcification on all equipment according to manufacturers recommendation will be carried out.

Proper documentation of all equipment records and services to be maintained and entered into the systems equipment database. **(See Tables 3 & 4).**

Technicians will only carry out controlled modification of equipment as directed by manufacturer. Protective clothing to be worn when appropriate i.e. gloves, goggles and protective shoes.

The appointment of a safety officer should be considered. (**Refer to Appendix A2**). His duties may include:

1. Detailed programmes designed to reduce accident, especially in the workshop area, including patient and staff facilities.

2. Accurate records and periodic reviews of those records must be kept. Since these may be required by law, or by the organisation.

3. The safety officer should seek to provide appropriate training facilities for staff and encourage them to observe and report routinely any conditions that may endanger safety.

Local rules may include:

A protocol for fluid spillage. Floors must be mopped immediately to prevent accidents.

A) Transport of specimens to the laboratory. Correct identification of specimen bottles is essential.

Extra care and precaution is needed when dealing with electrical equipment. Never handle electrical equipment with wet hands. Never change a fuse without first disconnecting the electrical supply. All major and minor faults must be reported. (**See Appendix B**).

Disinfecting agents can be harmful when they come into contact with skin or eyes. Wash affected area with plenty of running water. See a doctor if necessary.

Caution must be exercised at all times to ensure that equipment and boxes or containers are stored safely and that fire exit routes are kept free.

All equipment must be unplugged when not in use. Fire extinguishers to be checked regularly by trained staff. Do all members of staff know of the location and be able to use them correctly. If smoke detectors are installed, periodic maintenance by a qualified person is essential.

Electrical leads to be checked periodically. Frayed ones removed from equipment and new ones fitted.

Earth leakage checks are essential. **(Appendix 'B')**

<u>ACCEPTANCE TEST LOG (FOR ALL TYPES OF EQUIPMENT)</u>

Equipment Type
Order No.

Serial No.

Plant No.
Accessories
Cost

Manufacturer
Phone Contact
Supplier

Date of Acceptance
Warranty

Signature of authorised person

Location or technical document

Received by:

Table 1

ACCEPTANCE CHECKS AS DEFINED IN ACCEPTANCE TESTING DOCUMENT

Remarks – fail - corrected – pass – faults or n/a

General
a. Packing
b. Equipment intact
c. As specified
d. Accessories
e. Documentation

Technical
a. Equipment complete
b. Equipment undamaged
c. Knobs, fuses intact
d. Liquids correct
e. Wheels 7 castors

Markings
a. Class 1 or 2
b. Type B, BF, CF
c. Type AP, APG

Mains connection
a. Cable, plug intact
29 b. Plug connection
c. Cable colour code
d. Cord grip, fuse
e. Equipment protection
f. Fuse rating
g. Voltage setting
h. Earth terminal symbol

Electrical
a. Electrical measurement
b. Earthing of equipment
c. Earthing of accessories
d. Insulation resistance
e. Earth leakage current
f. Enclosure leakage current
g. Patient leakage current

Paragraph 3 (Installation)
1. Equipment function correctly
User tests, brief tests
Comprehensive tests
Supplier Test

---

a. Electrical connections
b. Plug/sockets
c. Permanently Installed
d. Other services
e. Environmental test
f. Controls
g. Indicators function
h. Alarms function
i. Emergency stop

Paragraph 4 (Formal Acceptance)

4.1 Equipment accepted
Supplier informed
Supplies officer informed
Equipment to User

Further Action

Required

4.2 Modification made by

Supplier and agreed
    Pass Fail

4.3 SERIOUS SHORTCOMINGS REMAIN

DEFECT ACTION REQUIRED

RELEVANT HEALTH DEPT
INFORMED
    HN (Hazard)
No.

Table 2

MACHINE SERVICE RECORD

MAKE
DATE

MODEL
LOCATION

NUMBER
HOURS

<u>Parts to checked and replaced</u>

1. Plug cable
2. Pumps and gears
3. Motors and inserts
4. Pump and hansen connector O-rings
5. Dialysate lines
6. Internal and external filters
7. Suction probes
8. Pressure isolators
9. Water and drain lines

10. All connetors.
11. Air filters
12. Earth continuity
13. Earth leakage

<u>Parts to be checked and calibrated</u>

1. Conductivity
2. Temperature
3. Transducer
4. Air detector
5. Priming detect.
6. BLD
7. R.O. output

Machine should be: -

1. Heat disnifected
2. Chemical disinfected
3. Citric cleaned

Table 3

After every service, check the conductivity, temperature, flow rate, transducers and everything recommended by the manufacturer.

The above service record is mainly for a dialysis machine. It can be adapted it suit various requirement.

TRAINING AND PRACTICE

Technical and scientific teams will provide basic instruction on equipment and machine theory to those personnel interested. In a hospital environment, nurses and doctors may show interest. The chief technician may want to keep a record of all personnel receiving any appropriate instruction including their status within the organisation.

Technicians will teach patients and customers the basic theory concerning the equipment they frequently use. This basic theory should be itemised and patients/customers should be required to sign when completed.

In a dialysis unit, water treatment theory will be introduced at an early stage. This is an essential part of the training, as many people are not aware of the various components of the water system and the impurities which are harmful to chronic renal patients. Technicians will attend training courses recommended by the manufacturer and by the organisation.

The status of the chief engineer or technician should attract enough favourable points, which should be awarded for budgetary responsibility, the specialised nature of the work, and additional managerial responsibility. He may therefore be the designated training officer.

(See Appendix A1).

Completed service reports for all types of equipment should be maintained and filed away in a safe place. Such an important document maybe useful in a coroner's court.

All members of staff should be given the appropriate training in order to report and successfully log every single fault that develops on all types of equipment. This saves time and money.

An appropriate training officer is strongly recommended for the training of all staff members. **(Appendix A2)**

MACHINE SERVICE REPORT

MAKE   DATE

MODEL   LOCATION

NUMBER   HOURS

JOB DETAILS
_____

_____

_____

_____

_____

WORK CARRIED OUT
_____

_____

_____

_____

_____

_____

_____

Check and or calibrate

CONDUCTIVITY	
FLOW RATE	
TEMPERATURE	FLOW RATE
CONDUTIVITY	SAMPLE TAKEN

CITRIC RINSED	CHEMICAL CLEAN
R.O. OUTPUT	R.O. CLEANED

Table. 4

FAULT SHEET

EQUIPMENT	NAME	MODEL	LOCATION

REPORTED BY		DATE

FAULT

COMMENTS

_____

_____

_____

_____

_____

REPAIRED	BY DATE

NAME   SIGNATURE

Table 5

APPEARANCE AND ENTHUSIASM

Technicians and service engineers will endeavour to be neat, tidy and have an air of confidence about them. Technicians and service engineers will not attend any service calls whilst under the influence of alcohol.

All members of the technical staff will show enthusiasm towards their work, given that their work ranges from the mundane to the highly technical and from the known to the unknown.

CONFIDENCE AND ASSERTIVENESS

All members of staff will seek to acquire knowledge in order to enhance their position.

In-house-training; the staff should be encouraged to make continual contact with the training and development department of the organisation.

Information and knowledge acquired will be transferred to other member of the working team.

The occasional presentation by a team member of staff on a chosen and appropriate work related subject should be encouraged and given.

WORKING PRACTICE

Technicians and all associates will aim for solidarity within the group.

We will seek to work on common ground in a democratic manner.

In the case of a merger of two separate companies, cross technical cover to commence initially by gathering as much technical and other information on different types of equipment and environment as possible.

This is vital for the essential smooth running of the system.

BRIEFING AND DEBRIEFING

Regular meetings with operations manager to discuss organisational, local issues, news and rules.

Technician's and engineer's work performance and loyalty to the organisation will be monitored by an appropriate manager, i.e. through regular appraisals.

Regular technical meetings should be held at predetermined times and venue, if possible, to sort out any technical or non-technical problems.

It is important that meetings are held in an open, friendly manner.

Briefing and debriefing creates a feed back loop which is vital to group dynamics. Members are able to focus on their own behaviour and that of other group members including interaction between group members. Communication is enhanced and it facilitates positive working relationships and activities.

This may help people to introspectively create a feedback arena where new values, attitude and awareness are considered and evaluated. This creates an atmosphere where there may be total co-operation and co-ordination of effort from group members.

An element of sensitivity comes into play and people appreciate the needs and feelings of others which lead to the acceptance, tolerance of and being considerate of other members including suggestions, ideas and new information.

In any organisation, the working skill, behaviour, attitude and awareness of individuals as well as group members are important factors in developing the objectives of the organisation.

Therefore the process of feedback developed during briefing and debriefing between members is vital when training methods, aimed at increasing group and individual effectiveness are being considered and used.

<u>STRUCTUAL ILLUSTRATION</u>

```
┌─────────────────┐          ┌─────────────────┐
│   (Verbal)      │          │    Group        │
│  Communication  │ ◄──────► │  Discussion     │
│   (Written)     │          │                 │
└─────────────────┘          └─────────────────┘
                                      │
                                      ▼
┌─────────────────┐          ┌─────────────────┐
│  Observation    │          │   Feedback      │
│     By          │ ◄─────── │    From         │
│   Manager       │          │    All          │
│                 │          │  Participants   │
└─────────────────┘          └─────────────────┘
                                      │
                                      ▼
                             ┌─────────────────┐
                             │    Advice       │
                             │    From         │
                             │    Group        │
                             │   Members       │
                             └─────────────────┘
                                      │
                                      ▼
                             ┌─────────────────┐
                             │  Assessment     │
                             │     By          │
                             │   Manager       │
                             └─────────────────┘
```

Figure 4
Input from the organisation with respect to local rules and protocol is essential.

Your Body Language

18. BODY LANGUAGE

Body language tells everything about a person, apart from the words that they use. This includes how you compose yourself. Posture, gesture, mannerisms, facial expressions, tone of voice, eye contact and how you meet customers are all very important.

Body language can be classified under communications. It expresses how we feel, but do not say. It can portray the negative or positive aspects of our verbal communication skills consciously or unconsciously.

We must therefore make conscious effort to correct any defects in our body language. Whether it is intrinsic or acquired from external influence.

Understanding our own body language and that of others can be very important to our communication skills.

In my opinion a nice smile can be infectious and travel many miles.

The vast majority of my compatriots seem to have some difficulty in smiling when meeting a customer and providing him or her with that vital service.

Many Africans and other Third World people including some members of some developed countries have intimated this fact; and in many circles especially African, it has always been a topic of discussion.

Smiling and being affable to customers is part of body language. It costs nothing.

Not smiling is to show a sign of antipathy, disregard and ingratitude, shows ignorance and can be costly in terms of customer level and economic. We cannot afford not to smile.

Learn to open a harmonious conversation. This helps prevent 'distancy' from keeping you and your customers apart.

Education

19. THREE EDUCATION DOMAINS

Technicians and engineers seem to work and learn from the known to the unknown. We must therefore be made aware of the necessary development of our educational domains.

To develop the necessary skills in order to provide scientific and technical support to any organisation and our customers, we need to develop our hand and technical skills in the workshop, the running of the equipment and assembling of parts; our attention, our perception and response. The area to be developed concerns our 'PSYCHOMOTOR DOMAIN'.

Our behavioural intention concerns the information we acquire and the knowledge we pick up. This is our 'COGNITIVE DOMAIN' which essentially incorporates:

A)   Gaining knowledge

B)   understanding and thinking

C)   The application of the knowledge gained.

This is the mind's processing of information and hence the correct application of our technical skills and routines.

Our feelings, emotions, value and attitude and awareness is of vital importance in creating a favourable public reaction. Hence the final domain to be considered and developed in order to positively improve and enhance our attitude, values and emotions is our 'AFFECTIVE DOMAIN'.

To instil a positive attitude in someone we must remember the three 'E's':

A)   EXAMPLE: - which can be demonstrated

B)   EDUCATION: - which can be achieved through studying

C)   ENFORCEMENT:- of rules and protocol if necessary

In our (technicians and engineers) field, attitude relates to courtesy, empathy, and consideration of our customers and members of the organisation.

We must not stoop below the level of a high moral value and etiquette and accept bribes in order to perform our duties.

The

Theories

20. LEARNING THEORIES

As most technicians and maybe some engineers tend to reach a learning plateau quite often too easily, I would suggest the introduction in some form of learning theory and method (which may not be totally technical) into the system.

There are several ways of learning. For example watching, listening, participating and studying.

Many factors can also affect our learning. For example; sex, age, motivation, concentration span, memory and instructional manner.

To change our attitude we need to incorporate some learning aids; i.e. develop our mental skills, physical skills and the way we handle our emotions.

It is generally accepted that learning skills can be divided into three domains, i.e. skill learning which is the co-ordinating of mind and muscle and can combine to achieve normal proficiency. This is our PSYCHOMOTOR DOMAIN.

There are a number of explanations about why and how people learn. People learn because they may want to acquire knowledge about themselves and others, animate or inanimate.

The two most common learning theories are 'COGNITIVE' and 'BEHAVIOURIST'. Cognition is concerned with the information and knowledge we acquire and relates to our mental process such as perception and thinking and the application of that knowledge.

The behavioural theory is associated with; human relationships and style of relationships. We must develop concern for people and production in order to positively and efficiently serve customer and nation.

Learning may involve the change of behaviour by the process of conditioning and can be achieved by:

A)   STIMULUS: - when a correct response is rewarded with the approval of the instructor or manager.

B)   RESPONSE: - where incorrect response is identified and corrected.

C)   REINFORCEMENT: - i.e. a positive and correct response is identified and the appropriate award given.

As the writer B.F. Skinner explains:

A)   'speak' of specific achievement.

B)   'Give bonus or pat on the back'.

C)   'Make task achievable'.

Frank Sikapa

Motivation

21. MOTIVATION AND REWARD

There are a number of underlying assumptions that motivate a man or woman. The significance of which provides the basis for the design of an effective reward system. A significant system whereby institutions, governments and organisations can attempt to influence and control the behaviour of their employees.

Members of the Third World countries are too biased in favour of economic incentives as the only means of motivation in order to maximise their work output and enhance public relations. Pure greed is another factor.

We (technicians and engineers) should be aware that there are other needs just as important as economics.

Such needs advocated by the main exponents of 'motivation' (Hertzberg, Maslow, McGregor and Mayo) are ; status, power,

Frank Sikapa

appreciation, achievement, benefits, involvement, decision making,

recognition of achievement and control over work.

HOW CAN OUR LEADERS USE THE MOTIVATING FACTORS TO MOTIVATE TECHNICIANS

AND ENGINEERS?

Recognise that individuals have their own set of needs.

Recognise that needs can change over time and according to circumstance.

Such needs will vary in accordance to our early life, experience, heredity and education.

Recognise that we need good basic working conditions in order to self-actualise. (Maslow).

Each employee needs recognition feedback and a sense of belonging.

Be aware of leader importance of motivating individuals according to their own particular need and want.

Recognise the value of good communication.

HYGIENE FACTORS (MAZLOW) SIMPLIFIED (MOTIVATING TEAM OF ENGINEERS/TECHNICIANS)

COMPANY POLICY AND ADMINISTRATION

Availability of clearly defined policies, adequacy of communication, efficiency of organisation, degree of 'red tape'.

SUPERVISION

Accessibility, competence and personality of manager or supervisor.

INTERPERSONAL RELATIONSHIPS

The relations with supervisors, subordinates and colleagues. The quality of social life at work.

SALARY

The total rewards package, such as salary, pension, company car and other 'perks'.

STATUS

A persons position or rank in relation to others, symbolised by title, parking space, car, size of office, furnishings where appropriate.

JOB SECURITY

Freedom from insecurity, such as loss of position or loss of employment altogether.

PERSONAL LIFE

The effect of a person's work on family life, e.g. stress, unsociable hours or moving house.

WORKING CONDITIONS

The physical environment in which work is carried out, the degree of discomfort it causes.

Frank Sikapa

THE CYCLE OF MOTIVATION

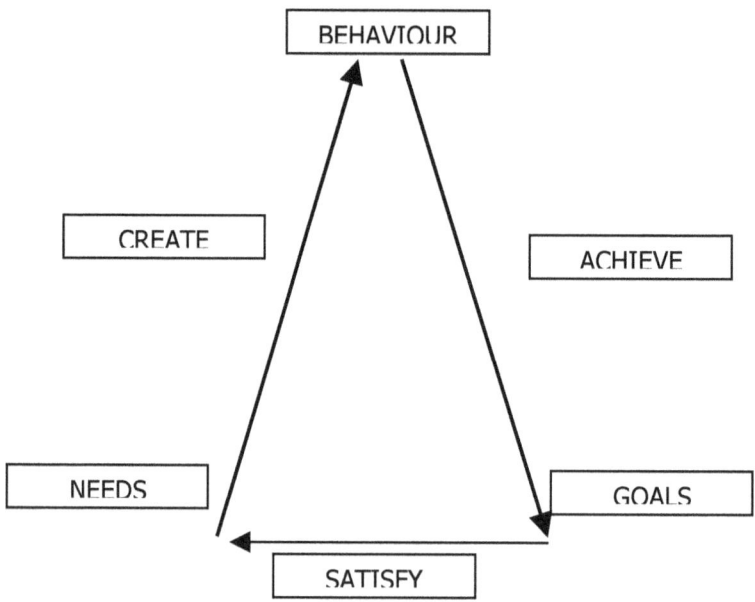

Figure 5

The key concept that underlies such consideration of motivation and reward system is that of the psychological effect it has on employees. This implies that individuals have a variety of expectations of the organisation and vice versa. These expectations include a whole pattern of rights, privileges and obligations between employees and organisation.

Applying the correct motivation factors may result in increasing the performance of the organisation.

There should be emphasis on co-operation rather that conflict, on integration rather than opposition. This may be achieved by integrating the needs of the individual to develop himself with the requirements of the organisation for effective task performance to the benefit of both.

Frank Sikapa

The Trade Union

22. TRADE UNION FORMATION AND MEMBERSHIP

This thesis has identified a number of significant methods and ways in which we as technicians and engineers can develop our skills, awareness and attitude towards others. Mainly, towards our customers.

One other channel through which industries can help to achieve our aims and objectives in this matter is by encouraging the formation of trade unions. In order to encourage good relations in the public and private sector as well as industrial relations nationally and internationally.

One of the oldest and most often quoted definitions of a trade union is that by Sidney and Beatrice Webb, 'a continuous association of wage earners for the purpose of maintaining or improving the condition of their working life'. (INDUSTRIAL DEMOCRACY. 1893).

Our working live is not influenced by economic, social, safety, physical, self esteem and self actualisation needs alone: In my opinion, our working live is influenced in a significant way by our attitude and the level of awareness plus proper consideration of other.

Encouraged union membership may not be confined largely to skilled workers alone. Unskilled workers must be allowed to unionise successfully.

All governments of the Third world countries and trade unions (when formed) must aim at developing very close relationships, for progress and to realise the need to emphasise the usefulness in the ability to control its members. The knowledge gained can then be transferred.

Benefits

23. BENEFITS

Proper co-ordination of efforts among technical staff and directed towards the whole organisation.

Knowledge gained through positive attitude can be transferred to other people.

Reduce stress factors due to a more friendly and democratic working environment.

Effective working team towards efficiency.

Hopefully, satisfies customers, who in my case are our patients.

Augment the status of the technician and bring to light their duties, activities that are sometimes demeaned.

Happy managers, workers and organisation.

Extra responsibility can be taken without the fear of the unknown.

A more civilised work force and environment and nation with regards to customer service.

Enhance the standing and contribution of the profession in the interest of the Various Third World nations and to the benefit of society.

If successful, can enhance the economic growth of the Third World, the spirit and indigenous enterprise of the people.

There will be recognition of Third World engineers and technicians internationally.

Hopefully, all of us will stop being egocentric and aim towards being empathic and considerate of our customers.

Various Third World nations can develop their own safety philosophy and standards including the interpretation and for their own safety requirements, **(See appendix 'B').**

Difficulties

24. DIFFICULTIES TO OVERCOME

Different customs and practices among individuals and groups of people in different countries.

Working practices may differ, e.g. working from home for some companies as opposed to working from base within the company.

In a case where two, major companies merge, difficulty in initiating cross technical cover because of different types of equipment on either side and environment.

Language and communication problems especially in Third World countries. Negative attitude instead of a positive attitude.

ACCEPTANCE OF MANAGER OR MANAGEMENT TEAM BY GROUP MEMBERS

Tact and diplomacy is called for.

Group discussion is important to make members feel part of a team and not work in isolation. Even at national level.

A democratic, instead of autocratic or free rein leadership style should be encouraged.

Hopefully, the group will give power to the implementors or leader to have the necessary energy and creativity to successfully fulfil the needs and expectations.

POSITIONAL POWER OF IMPLEMENTOR OR MANAGER

It is hoped that governments of Third World countries, and institutions will give adequate powers to the heads of technical and engineering industries. To use the necessary resources to achieve the required and up-to-date educational standard in terms of skill, attitude and awareness in all members of staff to the benefit of industries and nations.

Again, the major problem facing development and the really crucial one is the problem of economic scale. The majority of Third World countries are too small in terms of capability, ability, know-how and economic growth to generate an educational programme without the appropriate help from their colonial masters.

The optimum zone of development for a Third World population in a vast continent like Africa, is for the governments of the different colonial empire of the entire continent to co-ordinate their efforts and resources and pursue necessary educational policies to encourage the development of positive attitude and awareness. The longer the delay, the more difficult the task becomes. This can be a major problem.

Frank Sikapa

Lessons

Learnt

25. LESSONS LEARNT

The need to be effective as an implementor to fulfil the conditions of the project requires patience, tack and diplomacy.

There has to be a reasonable amount of co-operation from group members, including educational institutions.

Ideas may be accepted and measured as good or bad.

Massive education of the members of the Third World at large is needed and must be considered vital. The idea of education may be started in schools and colleges.

The correct definition of tasks and ideas to be accepted by the majority of the population and a democratic atmosphere encouraged if necessary.

People should accept others ideas and suggestions, keeping an open mind, (See Chapter 9)

Of paramount importance is the legitimate encouragement given by leaders of various countries and organisations in order to implement the corrective actions. This power should be measured as strong or weak.

There is no room for dictatorship or an overbearing personality.

Not even a place for a military regeme.

Being dogmatic and flaunting one's knowledge means that the necessary cooperation from others is not favourable and can lead to indignation.

The majority of people were found to be egotist with the belief that everyone is motivated by self-interest; an example of typical human behaviour **(See Chapter 4.1)** which prevails throughout most Third World countries.

My Conclusion

CONCLUSION

The preface highlights what the author was trying to achieve and how to go about resolving such a difficult problem in a theoretical and more importantly in a practical way.

The project also shows that all engineers and technicians use a combination of all the aforementioned education spheres, which need developing, especially in a Third World country where I come from. The education spheres being:

A)   Psychomotor

B)   Cognitive and

C)   Affective domains

I have also discussed how much knowledge and learning can be applied and used in our working environment. The idea is very challenging, exciting and can be very rewarding. It is up to us (leaders in this field) in many Third World countries to draw up a national strategy, with the help and collaboration of internal and bilateral institutions, together with a degree of private and public support. Such an advantage can be used to the best of our ability.

We may not be able to emulate the performance, powers and know how of the Western and developed world, but this can be a great giant leap towards achieving tangible goals and objectives.

There are three alternatives open to us all i.e., governments of the Third World and institutions such as (Institution of Diagnostics Engineers) of the developed nations. Firstly, to help enhance the profile of technicians and engineers. Secondly, to turn a blind eye to the disruption of our moral entity through bribery and corruption, and thirdly to disintegrate and disregard the positive ideas and efforts being practised by the developed, Western world.

It should be a comforting fact in the end, to observe that we have cleared a major obstacle, in that people of the Third World countries would except the ideas and learning theories proposed in this project.

With the major educational hurdles surmounted, what is left now is chiefly a question of strategy the intensity and earnestness of our demand. We must do our best.

W. Edwards once said; 'Do you know that doing your best is not good enough? You have to know what to do, and then do your best'.

Frank Sikapa

APPENDIX A1

JOB DESCRIPTION

JOB TITLE: CHIEF TECHNICAL TRAINING OFFICER
SALARY: NEGOTIABLE
ANNUAL LEAVE: ON ENTRY 0 – 20 DAYS

RESPONSIBLE TO: - DIRECTOR OF MACHANICAL ENGINEERING

1. THE HOLDER OF THIS POST IS A MEMBER OF A TECHNICAL TEAM PROVIDING TECHNICAL AND SCIENTIFIC SUPPORT FOR THE DEPARTMENT AND OTHER MECHINICAL AND ELECTRO MECHANICIAL EQUIPMENT.

PRINCIPAL DUTIES:

2. TO MANAGE A TEAM OF TECHNICIANS AND DELIVER SKILLED PERFORMANCE OF A BROAD RANGE OF WORK OR ACTIVITIES, THE GREAT MAJORITY OF WHICH ARE COMPLEX AND NON-ROUTINE.

3. TO PROGRAMME, TOGETHER WITH THE DEPARTMENT BUDGET HOLDER, THE REPLACEMENT OF OBSOLETE AND DEFECTIVE EQUIPMENT FOR THE DEPARTMENTS.

4. TO MONITOR NEW DEVELOPMENT IN ANY ASSOCIATED EQUIPMENT AND TO INITIATE EVALUATION PROGRAMMES AS REQUIRED TO SELECT THE MOST SUITABLE FOR THE DEPARTMENT.

5. TO DIRECT A PROGRAMME OF 'PLANNED PREVENTATIVE MAINTENANCE' FOR ALL EQUIPMENT, AND TO ALLOCATE DAILY WORK SCHEDULES.

6. TO INVESTIGATE DIFFICULTIES WHICH ARE THOUGHT TO BE EQUIPMENT RELATED, TO RECOMMEND CHANGES IN PROCEDURES, AND INTERFACE WITH MANUFACTURERS TO BRING ABOUT CHANGES IN THE EQUIPMENT.

7. TO ESTABLISH AND MAINTAIN A SYSTEM OF DOCUMENTATION FOR RECORDS, TOGETHER WITH AN INVENTORY OF EQUIPMENT DETAILS AND LOCATION, COMPUTERISED WHEREVER POSSIBLE.

8. TO ESTABLISH AND MAINTAIN A LIBRARY OF WORKSHOP MANUELS AND MAINTENANCE SCHEDULES FOR ALL EQUIPMENT IN USE.

9. TO ARRANGE FOR ALL TECHNICIANS TO ATTEND MANUFACTURERS RESIDENTIAL TRAING COURSES, AS REQUIRED, TO FULLY UNDERSTAND THE CONSTRUCTION, OPERATION AND MAINTENANCE OF THEIR EQUIPMENT.

10. TO PROVIDE ANY EQUIPMENT-RELATED TRAINING FOR STAFF AS REQUIRED.

11. TO MAINTAIN AN ADEQUATE LEVEL OF SPARE PARTS AND TEST EQUIPMENT, AND TO SEEK THE BEST SOURCES OF SUPPLY TO MAKE TO BEST USE OF THE ALLOCATED BUDGET.

12. TO SELECT NEW STAFF WHEN REQUIRED AND PROVIDE APPROPRIATE SPECIALIST TRAINING TO EQUIP THEM FOR THE WORK THEY WILL BE EXPECTED TO PERFORM.

13. TO PROVIDE TRAINING TO OVERSEA STUDENTS SEEKING WORK EXPERIENCE AS REQUIRED.

14. TO ESTABLISH AND MAINTAIN STANDARDS OF SAFETY WITH COORDINATION FROM THE SAFETY OFFICER.

15. TO BE ABLE TO ADAPT TO MEET ANY OTHER REASONABLE REQUIREMENTS THAT MAYBE APPARENT.

FURTHER INFORMATION

1. THE JOB REQUIRES SOME KNOWLEDGE OF ANALOGUE AND DIGITAL ELECTRONICS, AND PREFERENCE WILL BE GIVEN TO THOSE WITH PROVEN PRATICAL AND FRONT-LINE MANAGEMENT ABILITY.

2. CONSIDERABLE EMPHASIS IS PLACED ON THE POST HOLDERS ABILITY TO BE TACKFUL AND DIPLOMATIC IN SUPERVISION AND COMMUNICATION, NOT ONLY WITH OTHER TECHNICIANS BUT WITH OTHER MEMBERS OF STAFF.

3. THE POST HOLDER MUST AT ALL TIMES FULFILL HIS/HER REQUIREMENTS WITH DUE REGARD TO THE APPROPRIATE AUTHORITIES EQUAL OPPORTUNITIES POLICY.

4. THE POST HOLDER MUST AT ALL TIMES RESPECT CONFIDENTIALITY OF ELECTRONICALLY STORED DATA, IN LINE WITH THE REQUIREMENTS OF THE DATA PROTECTION ACT.

5. THE POST HOLDER MUST ENSURE THAT ALL DUTIES ARE CARRIED OUT WITH DUE REGARD TO THE POLICIES AND PROCEDURES OF THE GROUP; IN PARTICULAR, HEALTH AND SAFETY POLICIES.

6. THIS JOB DESCRIPTION IS INTENDED AS A GUIDE TO THE MAIN TASKS INVOLVED IN THE POST AND IS NOT INTENDED AS AN EXHAUSTIVE LIST OF DUTIES AND RESPONSIBILITIES. IT WILL BE SUBJECT TO AMENDMENT AND MAY BE CHANGED FROM TIME TO TIME AFTER CONSULTATION WITH THE POST HOLDER.

7. THE POST HOLDER MAY BE REQUIRED TO UNDERTAKE ANY DUTY APPROPIATE TO THEIR GRADE AT THE DISCRETION OF THE MANAGER.

Frank Sikapa

APPENDIX A2

JOB DESCRIPTION

JOB TITLE: SAFEFY OFFICER
GRADE: S.O. IV
SALARY: NEGOTIABLE
ANNUAL LEAVE: ENTRY – 20 DAYS
RESPONSIBLE TO: CHIEF TECHNICAL TRAINING OFFICER

1. THE HOLDER OF THIS POST IS A MEMBER OF A TEAM OF TECHNICIANS PROVIDING SUPPORT FOR ELECTROMECHANICAL AND MECHANICAL EQUIPMENT.

PRINCIPAL DUTIES:

2. TO ESTABLISH AND MAINTAIN STANDARDS OF HEALTH AND SAFETY WITH THE CO- ORDINATION FROM THE RISK MANAGEMENT AND CHIEF TECHNICAL OFFICER.

3. TO DRAW A PROGRAMME TO REDUCE ACCIDENTS IN THE DEPARTMENT.

4. TO KEEP ACCURATE RECORDS OF ALL ACCIDENTS AND RECORDS TO BE REVIEWED PERIODICALLY.

5. TO ENSURE FIRE EXTINGUISHERS ARE IN GOOD WORKING ORDER AND CHECKED PERIODICALLY.

6. TO ARRANGE AND PROVIDE TRAINING FACILLITIES FOR STAFF AND ENCOURAGE THEM TO REPORT AND DOCUMENT CONDITIONS WHICH MAY ENDANGER SAFETY.

7. TO INFORM RISK MANAGEMENT ABOUT MAINTAINENCE COSTS, THE COST OF BREAKDOWNS, REPLACEMENT COSTS AND EFFICIENCY OF THE SYSTEM.

8. TO LIASE WITH CHIEF TECHNICAL TRAINING OFFICER TO PROGRAMME LOCAL HEALTH AND SAFETY RULE AND PROTOCOL.

9. TO INITIATE A PROGRAMME OF AWRENESS WITH REGARD TO ALL EMPLOYEES DUTIES UNDER THE HEALTH AND SAFETY ACT.

10. TO BE ABLE TO ADAPT TO MEET ANY OTHER RESONABLE REQUIREMENTS THAT MAY BE APPARENT.

11. TO ESTABLISH AND MAINTAIN PROPER DOCUMENTATION FOR RECORDS, TOGETHER WITH AN INVENTORY OF EQUIPMENT AND LOCATIONS.

FURTHER INFORMATION

1. THE JOB REQUIRES DETAILED KNOWLEDGE OF THE HEALTH AND SAFETY AT WORK ACT.

2. EMPHASIS IS PLACED ON THE POST HOLDERS ABILITY TO MAINTAIN GOOD INTERPERSONAL RELATIONSHIPS WITH ALL MEMBERS OF STAFF AND SHOW LEVEL HEADEDNESS IN DANGEROUS SITUATIONS.

3. THE POST HOLDER MUST AT ALL TIMES RESPECT CONFIDENTIALITY OF STORED DATA, IN LINE WITH THE REQUIREMNETS OF THE DATA PROTECTION ACT.

4. THE POST HOLDER MUST ENSURE THAT ALL DUTIES ARE CARRIED OUT WITH DUE REGARD TO THE POLICIES AND PROCEDURES OF THE GROUP. IN PARTICULAR, HEALTH AND SAFETY POLICIES.

5. THIS JOB DESCRIPTION IS INTENDED AS A GUIDE TO THE MAIN TASKS INVOLVED IN THE POST AND IS NOT INTENDED AS AN EXHAUSTIVE LIST OF DUTIES AND RESPONSIBLITIES. IT WILL BE SUBJECT TO AMENDMENT AND MAY BE CHANGED FROM TIME TO TIME AFTER CONSULTATION WITH THE POST HOLDER.

6. THE POST HOLDER MAY BE REQUIRED TO UNDERTAKE ANY DUTY APPROPRIATE TO THEIR GRADE AT THE DISCRETION OF THE MANAGER.

APPENDIX B

THE SAFETY PHILOSOPHY

1. British Standards (BS) interprets the general safety requirements in relation to the type of equipment used in the United Kingdom.

2. Sometimes the safety philosophy relates to a 'safety system' rather than separate pieces of equipment considered in isolation. An example of this would be a 'safe haemodialysis treatment'.

3. As the patient is earthed to the equipment via a conductive fluid PATH during haemodialysis, the patient leakage current and the earth leakage current are, under single fault conditions, the same. To ensure patient safety, the dialysing fluid outlet in the equipment must be effectively and reliably earthed.

Frank Sikapa

Appendix C

Form of Indemnity. A. (Equipment on loan for trial or testing)

ANAGREEMENT made

day of   month   year

BETWEEN THE_____**(the Supplier)**

WHEREAS: -

1. The supplier is the owner of the equipment described in the Schedule (the equipment)

2. The supplier wishes the authority to use the equipment for the benefit of the supplier for the purpose of evaluation, testing, research, design investigation or trial.

    IT IS HEREBY AGREED: that the supplier shall lend and the authority shall borrow and use free of charge the equipment for the period specified in the schedule in the premises specified in the schedule (the Premises) on the terms set out below.

1. The loan of the equipment shall be deemed to be a contract for the hire of goods as defined by section 6 of the Supply of Goods and Services Act (year).

2. The supplier shall be liable for and shall indemnify the authority and the Secretary of state for health against all liability in respect of personal injury to or the health of any person, loss or damage to property and any loss or expense in consequence of or in any way arising out of the installation, presence, use or removal of the Equipment on or from the Premises provided that this indemnity shall not

extend to liability resulting from the negligence of the Authority's own servants or agents.

4. The Supplier shall insure against its liability under condition certain conditions in respect of any one incident.

5. The Supplier upon request shall produce to the authority documentary evidence that the insurance is properly maintained.

6. Should the Supplier default in insuring the Authority may itself affect insurance and may charge the cost together with an administrative charge of certain percentage to the supplier.

7. The supplier shall provide the Authority with written evidence on the safety of the Equipment drawing attention to comply with relevant British (or country of origin) Standard or Department of Health (DoH) specifications or aspects of safety that have not been fully tested. Restrictions on the use of the Equipment necessary to ensure the safety of patients or staff shall be pointed out to the Authority.

8. A delivery note shall accompany the delivery of the Equipment identifying the Equipment by serial number or otherwise.

9. Detailed instructions in the use of the Equipment shall be given to the Authority's nominated staff by a qualified agent of the Supplier and detailed instructional manuals where available shall be supplied to the Authority.

10. Upon receipt of a written request at any time from the Authority the Supplier shall remove with all practicable speed free of charge at that time provided the Authority with receipt of the Equipment.

11. The Authority shall permit the Supplier to remove the Equipment from the Premises on receipt of a reasonable notice in writing.

12. The Supplier shall be responsible for the cost of reinstating the Premises including the services therein to the satisfaction of the Authority.

13. The Equipment shall remain continuously at the Supplier's risk during and after the period of loan.

SIGNED on behalf of the
    Authority_____Print_____

SIGNED on behalf of the
    Supplier_____Print_____

THE SCHEDULE

1. The Equipment.

   Model/Mark No.

   Serial No.

   Value.

   Description.

2. Period of Loan

   Years   months commencing the   day of   year

   The Premises

I hope you have fully enjoyed your Reading.

Good luck and God Bless.

Frank Sikapa (MSc., P. Eng., F. I. Diag.E.)

Now is the time to test you knowledge and power of retention.

Please answer the questions below.

Frank Sikapa

Questions.

1. Examine the role of feedback in a technical control system.

2. Write briefly about Trade Union membership.

3. What are the three Educational Domains – Discuss.

4. Consider the role of the manager's function in controlling the technical staff and stock level.

5. What is the secret of good instructional technique.

6. Why is lesson planning important.

7. What are the main characteristics a technician must try and exercise.

8. How do we apply our technical and mental approach.

9. What is the role of self confidence and efficiency in problem solving.

10. Why is serving the customer can sometimes be more important than servicing the equipment.

11. Training is an important aspect of our technical work-Discuss

12. What is the role of a front line manager.

13. Discuss the method of stock keeping.

14. What are the objectives of a good maintenance regime.

15. What, are the key points to be considered when disposing of equipment.

16. Why is body language and technique important – Discuss.

17. Discuss the various learning theories including Motivation.

18. What are the Hygiene factors that any organization could apply to her employees.

19. What difficulties are there in all the educational domains.

20. Discuss the learning theories that technicians should apply.

21. Diacuss the benefits of positive organizational control and monitoring.

Bibliography

1. **Renal Dialysis Technologist Companion.** (Frank Sikapa)

2. **Gambro College Publication** (Gambro Ltd. UK)

3. **Fluid and Electrolytes** (Abbot Laboritories Inc.)

4. **Fluid Removal Management** (Cobe Hospal) UK

5. **Water Treatment for home dialysis**(Cobe/Hospal/Gambro)

5. **Organization of Maintenance** (Emil Woolf/Karam Singh)

6. **Textbook of Medical physiology** (A. C. Guyton)

Frank Sikapa

About the Author

Frank Sikapa MSc., PEng., FIDiagE
Retired Chief Technician
(Department of Renal Medicine & Transplantation)
St. Bartholomew's Hospital
London EC1A 7BE
United Kingdom

Frank Sikapa was the first Black person of African origin to have worked in the field of Dialysis in Great Britain and the whole of Europe. He, together with the pioneers, started the dialysis complex at St. Leonard's Hospital, London. He joined the team in August 1977.

He lectured Post-Graduate Students in Dialysis Servicing, Organization and maintenance of in-house and the outreach system for the Medical Electronics College of St. Bartholomew's Hospital, (London University), collaborating with The World Health Organization to provide Training, Scientific support, research and Development in Health Care Equipment.

Now, Frank has written the ultimate guide in technical skills, awareness, and positive attitude in his specialized field.

You will find energy, enthusiasm, expertise and genius in the pages of his book. A fascinating reading for any technically minded person and engineer.

His depth of knowledge and expertise is amazing and impressive.

The book is a pleasure to read, valuable and easy to resource.

www.ingramcontent.com/pod-product-compliance
Lightning Source LLC
Chambersburg PA
CBHW021944170526
45157CB00003B/918